Straight Lines, Parallel Lines, Perpendicular Lines

1
5
6
7
9

Straight Lines, Parallel Lines, Perpendicular Lines

By MANNIS CHAROSH
Illustrated by ENRICO ARNO

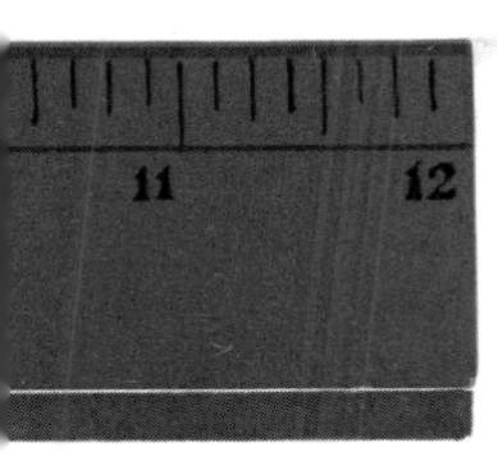

THOMAS Y. CROWELL COMPANY
NEW YORK

YOUNG MATH BOOKS
Edited by Dr. Max Beberman
Director of the Committee on School Mathematics Projects,
University of Illinois

Estimation
by Charles F. Linn

Straight Lines, Parallel Lines, Perpendicular Lines
by Mannis Charosh

Weighing and Balancing
by Jane Jonas Srivastava

What Is Symmetry?
by Mindel and Harry Sitomer

Published in Canada by Fitzhenry & Whiteside Limited, Toronto.

Manufactured in the United States of America

L.C. Card 76-106569

ISBN 0-690-77992-5 ISBN 0-690-77993-3 (LB)

4 5 6 7 8 9 10

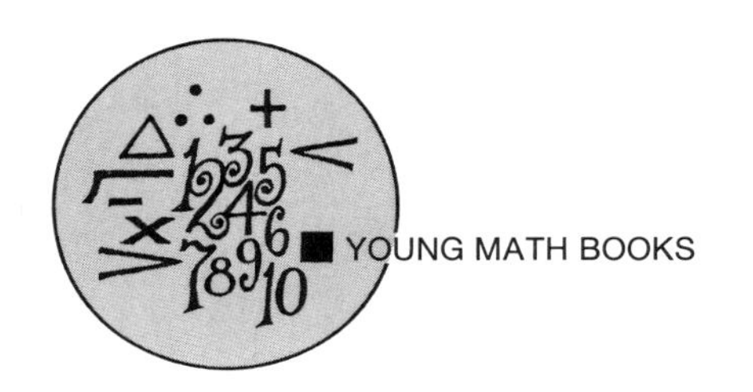

Straight Lines, Parallel Lines, Perpendicular Lines

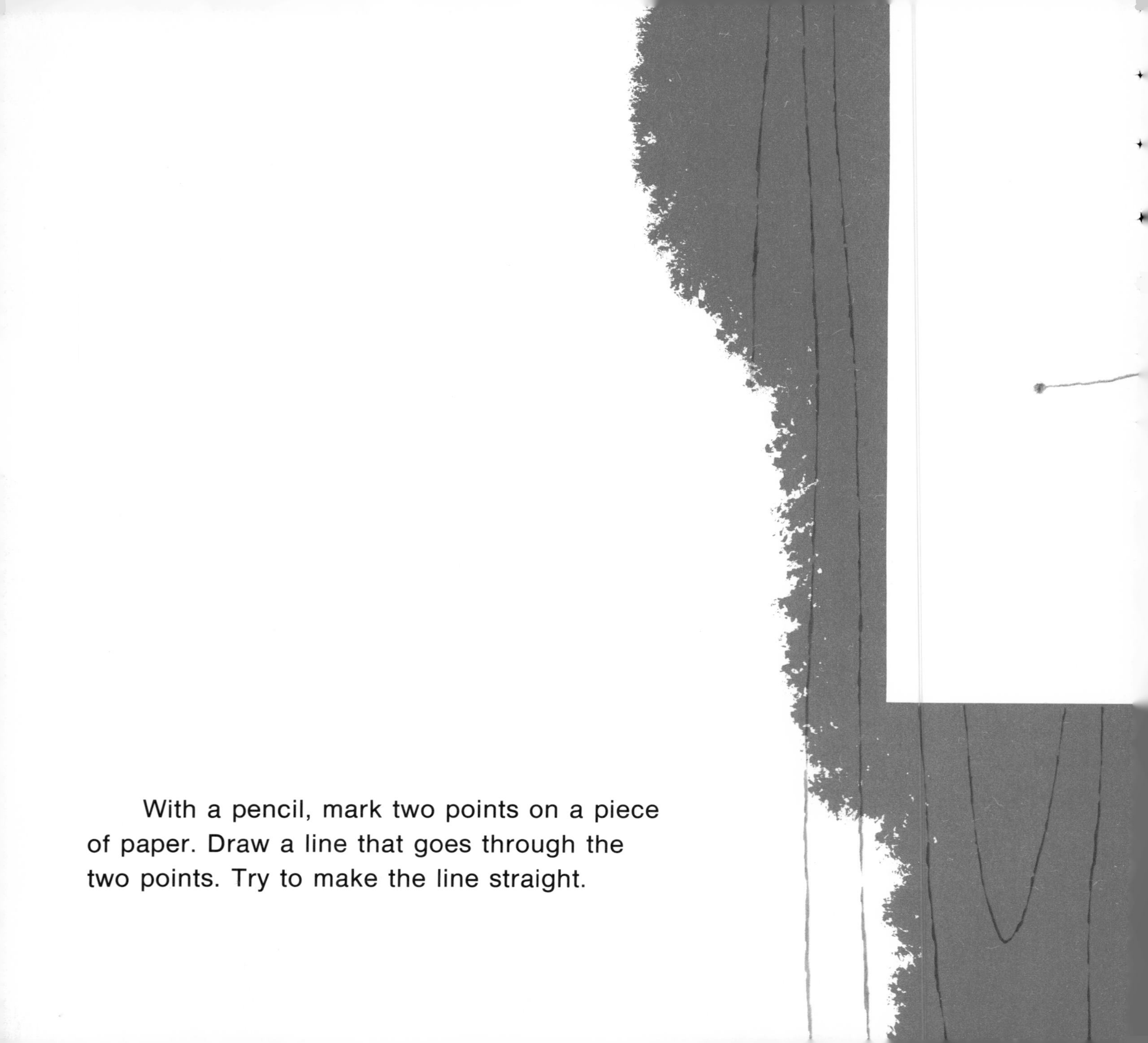

With a pencil, mark two points on a piece of paper. Draw a line that goes through the two points. Try to make the line straight.

1 4 5 6 7 8 9 10

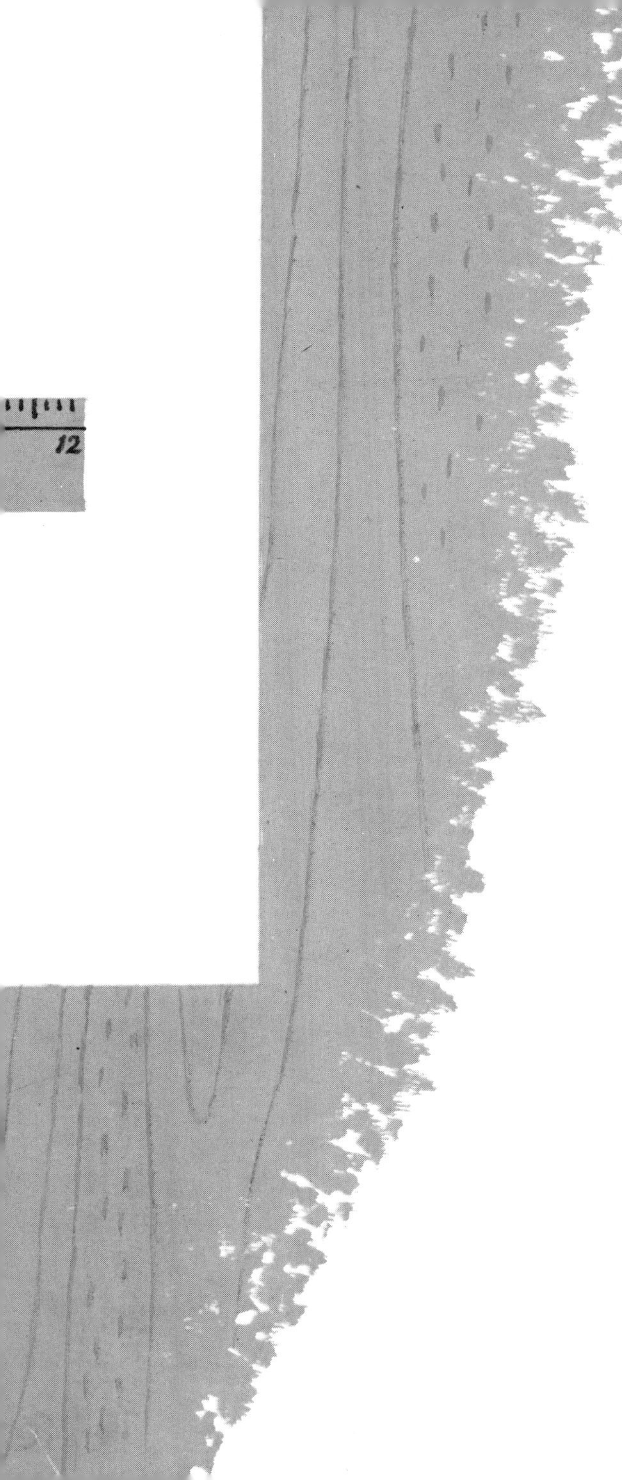

How can you tell whether it is a straight line? One way is to place a ruler so that its edge goes through the two points. If the line and the ruler's edge match, the line is straight.

Instead of the edge of a ruler, you might use the edge of a book, or the straight edge of a piece of school paper. You could use any edge that was made to be straight.

But suppose you did not have any of these objects.

How could you tell whether the line was straight?

If you pull the two ends of a piece of string apart tightly, its shape will be a straight line. Place the tightened string so that it goes through the two points marked on the paper. If the line you drew matches the string, the line is straight.

If all objects with straight edges were to disappear, you could still make a straight line. All you would need is a piece of string.

Are the edges of your kitchen table straight lines? You can test this by holding a tightened string alongside each edge. If the edges match the string, they are straight lines.

The edges of a chair are not always straight lines. Test the legs of the chairs in your house with a tightened string. Are there other edges that you are not sure of? Test them in the same way.

Can you guess with your eye whether three points lie on a straight line? Mark two points on a piece of paper. Pretend that a straight line has been drawn between these points. Do not draw the line. Mark a third point, which you think will lie on that line. Place the tightened string on the first two points. If the third point lies on the string, your guess was good. Whenever three or more points lie on the same straight line, we say that the points are COLLINEAR.

Did you ever notice that checkers on a checkerboard sometimes lie on a straight line? A checkerboard is like a classroom. A classroom may have five or six rows, with six or seven seats in each row. A checker-board has eight rows, with eight squares in each row. Suppose Francine sits in the second row–fourth seat in her classroom. When Francine goes to her seat, it is like placing a checker in the second row–fourth square of a checkerboard.

Place eight checkers in the second row from the left side of the checkerboard. Hold the tightened string so that it goes through the centers of the checkers. (You may have to move some of the checkers a little.) The string will show that the centers of the checkers are collinear. Let's call the straight line made with the string the LINE OF CENTERS of the checkers.

Many lines of centers can be made on a checkerboard. Here is one. Place checkers on the

first row–first square
third row–third square
fifth row–fifth square
seventh row–seventh square

You can now test with the tightened string to see if the centers of the checkers are collinear.

Four more checkers can be placed on the line of centers of the first four checkers. They would be placed on the second row–second square, the fourth row –fourth square, the sixth row–sixth square, and the eighth row–eighth square.

That was easy!

But there are other lines on the checkerboard that are harder to find.

Place checkers on the first row–second square and the fifth row–fourth square. Can you put a third checker between them so that all three centers are collinear? You can show with the tightened string that the first two checkers are collinear with a checker placed in the third row–third square. Where would a fourth checker be placed so that the centers of all four checkers were collinear?

Here is another line of centers you can make on a checkerboard.

Place checkers on the third row–seventh square and the fourth row–fourth square. Try to place a third checker so that all three centers will be collinear. Now test what you have done. The third checker should be on the fifth row–first square.

Try to find other sets of three checkers whose centers are collinear. Give the row and square for each of the checkers. Test each set.

Look at the top two lines of a page in your school notebook. These lines do not cross each other. If you drew both lines longer in both directions, they still would not cross each other. No matter how long you drew the lines, they would not cross each other.

They were made that way.

Such lines are called PARALLEL lines. Parallel lines may be close to each other—like the top two lines of your notebook page. Parallel lines may be far apart—like the top line and bottom line of your notebook page. These lines

are parallel because they are drawn so that they will not cross each other even if they are made longer.

Fold a piece of paper so that two opposite edges come together.

The crease will be parallel
to each of the two edges.

You can find many examples of parallel lines if you look about your house, your classroom, your street. In each of the following examples, judge with your eye that the lines are parallel.

In your house: the opposite edges of a rug.

In your classroom: the upper and lower edges of the chalkboard.

On your street: the cracks in the sidewalk.

You will find more examples of parallel lines in stores, in pictures, and in many other places.

You can also form parallel lines on the checkerboard. Here is an easy example.

Place a red checker on each square of the second row. Now place a black checker on each square of the fourth row. The line of centers of the red checkers is parallel to the line of centers of the black checkers.

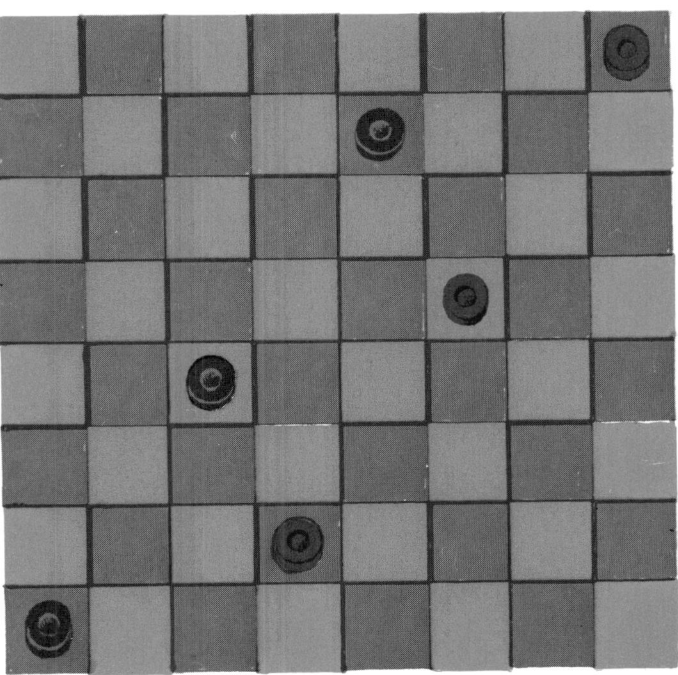

Here is another example of parallel lines on the checkerboard. Place black checkers on the

second row–fifth square
fourth row–third square
sixth row–first square

Now place red checkers on the

fourth row–seventh square
sixth row–fifth square
eighth row–third square

The line of centers of the red checkers is parallel to the line of centers of the black checkers.

Here is one more example. Place black checkers on the

first row–first square
third row–fourth square
fifth row–seventh square

Now place red checkers on the

fourth row–second square
sixth row–fifth square
eighth row–eighth square

The line of centers of the red checkers is parallel to the line of centers of the black checkers.

Try to form other pairs of parallel lines of centers.

Fold a piece of paper so that two opposite corners meet. The crease is a straight line.

Now open the paper and then fold it so that the two ends of the crease come together. When you open the paper again, you will see two lines that cross each other. The paper will be divided into four parts.

Place a book so that the edges of one corner fit along the lines that form one of these four parts. The corner of the book should be at the point where the lines cross. You will find that the corner of the book also fits into each of the other three parts. The corner of the book and the lines forming each of the four parts should match.

We say that the edges at the corner of the book are PERPENDICULAR. The creases in the paper are also perpendicular.

Judge with your eye whether or not the corner of a book will fit between these lines:

The corner of the book would have to be at the point where the lines meet, and the edges of the book would have to lie along the lines. If you think that the corner of the book would fit, you could say that the lines are perpendicular. But you probably agree that they will not fit. These lines are *not* perpendicular.

You can form perpendicular lines of centers on a checkerboard.

Place black checkers on the first square of each row.

Now place red checkers on all the squares of the third row except the first square.

The line of centers of the black checkers will be perpendicular to the line of centers of the red checkers. Test this by using the corner of a book.

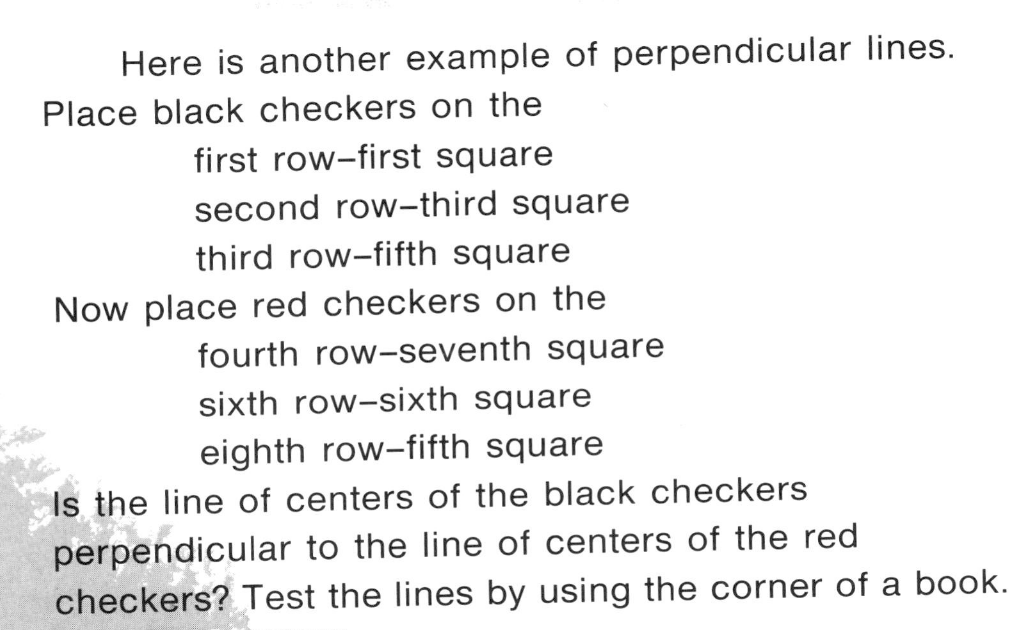

Here is another example of perpendicular lines.
Place black checkers on the

first row–first square
second row–third square
third row–fifth square

Now place red checkers on the

fourth row–seventh square
sixth row–sixth square
eighth row–fifth square

Is the line of centers of the black checkers perpendicular to the line of centers of the red checkers? Test the lines by using the corner of a book.

MAGAZINE

Perpendicular lines often appear in the same objects that have parallel lines.

In a rug the opposite edges are parallel, but the edges that meet are perpendicular. The same is true for a picture frame, a desk, a chalkboard, a magazine. Curbs of avenues are often parallel. So are curbs of side streets. But curbs of side streets are often perpendicular to curbs of avenues.

Whenever something appears often around us, it is convenient to have a name for it. STRAIGHT LINES, PARALLEL LINES, PERPENDICULAR LINES are names of familiar shapes.

Look around you. Do you see them?

ABOUT THE AUTHOR

Mannis Charosh has lived all his life in Brooklyn, New York. He has taught mathematics to high school students and has written books, filmstrips, and motion picture narrations about mathematics and teaching mathematics.

Mr. Charosh is a chess enthusiast. In addition to playing the game, he has composed many chess problems. It is this interest that gave him the idea of using the checkerboard in *Straight Lines, Parallel Lines, Perpendicular Lines* to introduce the concept of coordinate geometry.

ABOUT THE ILLUSTRATOR

Enrico Arno has had a distinguished career as an illustrator of children's books. He was born in Mannheim, Germany, and educated in Berlin. In 1940 he emigrated to Italy, where he worked for book publishers in Milan and later in Rome. Mr. Arno came to the United States in 1947. He and his wife live in Sea Cliff, New York.

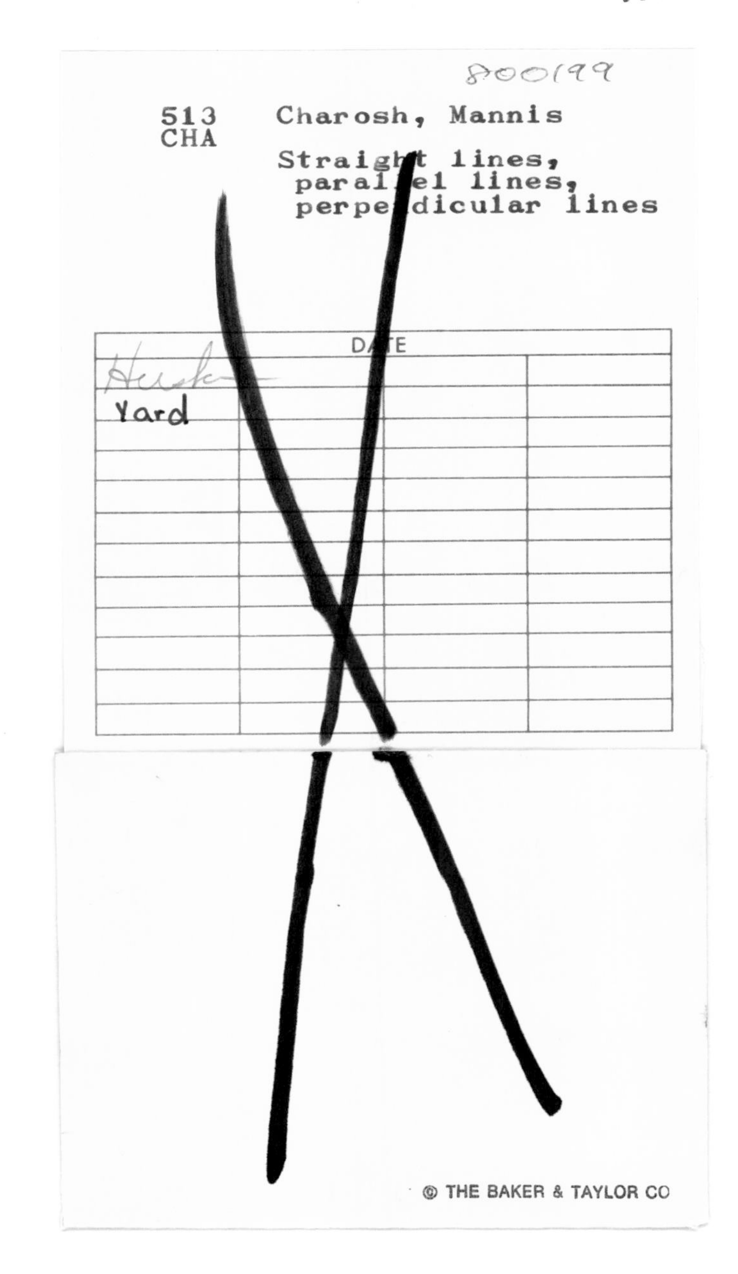
800199
513
CHA
Charosh, Mannis
Straight lines,
parallel lines,
perpendicular lines
DATE
Yard
© THE BAKER & TAYLOR CO